# HOW TO DOMINATE AS A VIRTUAL ASSISTANT

## BEGINNERS HANDBOOK

Nancy C. Paul

# Table of Contents

Title page

Preface

Welcome!

Hello there, I do not take it for granted that you are reading this book. There is tons of content out there but if you decide to consume this, rest assured that this will guide you and get you started if you take action.

This is the first step to your journey in virtual assistant, from Getting Started to the Services Offered, services Offered, it is safe to say this book is loaded. Make sure you read till the end, and make the most of it once done.

Much Love

**Nancy C. Paul**

## Introduction

As the world becomes increasingly digital, the demand for virtual assistants (VAs) is on the rise. A virtual assistant is someone who provides professional administrative, technical, or creative assistance to clients remotely, from a home office, or any other location with internet access.

The benefits of being a virtual assistant are numerous. For starters, you can work from anywhere, at any time, and enjoy a flexible schedule that allows you to balance work and personal life. You have the freedom to choose the clients and projects that interest you the most, and you can expand your skill set and learn new things as you work with different clients.

The market demand for virtual assistants is also on the rise. As businesses adapt to the changes brought about by the pandemic, many are turning to virtual assistants to help them stay organized, manage their online presence, and handle day-to-day tasks.

In this book, you will learn how to start, scale, and dominate as a VA. Whether you are new to the field or looking to take your virtual assistant business to the next level, this course will provide you with the knowledge, skills, and tools you need to thrive in the fast-paced world of virtual assistance.

Throughout this course, you will learn how to identify your skills and strengths, choose your niche and target market, create a business plan, set up your home office, offer a range of services, market and sell your services, manage your operations and finances, and overcome common challenges faced by virtual assistants.

By the end of this book, you will have the confidence and expertise to start your virtual assistant business, attract and retain clients, and build a successful career as a VA.

## Getting Started

Before you can start offering your services as a virtual assistant, there are a few key steps you need to take to ensure that you are prepared for the work ahead. This section will cover the following topics:

A. Identifying your skills and strengths
Before you can start your virtual assistant business, you need to identify the skills and strengths that you bring to the table. This will help you determine what services you can offer and what types of clients you are best suited to work with.

Consider the tasks you enjoy doing, the areas where you excel, and the skills you have acquired through your education, work experience, and personal interests. Some common skills and strengths that are in high demand for virtual assistants include:

Organizational skills
Time management
Attention to detail

Communication skills
Writing and Editing
Social media management
Customer service
Project management

B. Choosing your niche and target market

Once you have identified your skills and strengths, you need to choose a niche and target market for your virtual assistant business. This will help you focus your services and marketing efforts and make it easier to attract and retain clients.

Consider the types of clients and industries you are interested in working with, as well as the services you are most qualified to provide. Some popular niches for virtual assistants include:

Executive assistants
Social media managers
Customer service specialists
Content writers and editors

E-commerce assistants

Project managers

## C. Creating a business plan

To ensure the success of your virtual assistant business, you need to create a business plan that outlines your goals, target market, services, marketing strategies, and financial projections. This will help you stay organized, focused, and accountable as you start and grow your business.

Your business plan should include the following elements:

Executive summary

Market analysis

Services offered

Marketing and sales strategies

Financial projections

Management and operations plan

## D. Setting up your home office

One of the advantages of being a virtual assistant is that you can work from anywhere. However, to be productive and professional, you need to set up a dedicated home office that is conducive to work.

Your home office should be comfortable, quiet, and well-lit, with a desk, chair, computer, and other necessary equipment. You should also have a reliable internet connection, a phone, and a backup system for storing and backing up your files.

E. Legal considerations

Before you can start your virtual assistant business, you need to address any legal considerations, such as registering your business, obtaining the necessary licenses and permits, and ensuring that you are complying with tax laws and regulations.

You may also need to consider liability insurance and have contracts in place to protect yourself and your clients. It is recommended that you consult with a lawyer or accountant to ensure that you are complying with all legal requirements and protecting your business interests.

By following these steps, you will be well on your way to starting your virtual assistant business and building a successful career as a VA.

## Services Offered

As a virtual assistant, you can offer a wide range of services to clients, depending on your skills, strengths, and niche. Here are some common services offered by virtual assistants:

A. Administrative services
Administrative services are the most common services offered by virtual assistants. These services include:

Email management
Calendar management
Scheduling appointments
Travel arrangements
Data entry
Bookkeeping
Online research
Transcription

B. Technical services

Technical services are in high demand for clients who need help with their online presence and digital marketing. These services include:

Website design and maintenance
Search engine optimization (SEO)
Social media management
Email marketing
Graphic design
Video editing

C. Creative services
Creative services are in demand for clients who need help with their content creation and marketing.

These services include:
Writing and Editing
Blogging
Content marketing
Copywriting
Proofreading

Ghostwriting

D. Other services

Virtual assistants can also offer a variety of other services, depending on their skills and interests.

These services include:
Event planning
Project management
Customer service
Personal shopping
Translation and interpretation
Legal and paralegal services
When deciding which services to offer, it is important to focus on your strengths and interests, as well as the needs of your target market. You may want to consider offering a package of services that can be tailored to meet the specific needs of each client.

It is also important to communicate your services clearly to potential clients, through your website, social media profiles, and other marketing materials. Be specific about what services you offer, how you can help clients, and what sets you apart from other virtual assistants.

By offering a range of high-quality services that meet the needs of your target market, you can build a reputation as a skilled and reliable virtual assistant and attract more clients to your business

## Marketing and Sales

Marketing and sales are critical components of any successful virtual assistant business. Here are some strategies to help you attract and retain clients:

A. Define your target market

Before you start marketing your services, it is important to define your target market. Who are the clients you want to work with? What are their needs, challenges, and pain points? By defining your target market, you can tailor your marketing messages and strategies to reach the clients who are most likely to benefit from your services.

B. Create a professional website

Your website is your online storefront, and it should showcase your skills, services, and value proposition. Make sure your website is well-designed, easy to navigate, and mobile-friendly. Include a clear description of your services, your pricing, your contact information, and testimonials from satisfied clients. You may also want to include a blog or resources section to demonstrate your expertise and provide value to potential clients.

## C. Use social media

Social media can be a powerful tool for reaching potential clients and building your brand. Choose the platforms that your target market uses most frequently and post regular updates, including valuable content, industry news, and promotions. Engage with your followers by responding to comments, asking questions, and participating in industry-related discussions.

## D. Build your network

Networking is essential for any business owner, and virtual assistants are no exception. Attend industry conferences, join online forums and groups, and participate in local business associations. Build relationships with other virtual assistants, business owners, and potential clients. Offer to provide value and support to others in your network, and they will be more likely to refer clients to you in the future.

E. Offer value-added services

To stand out in a competitive market, consider offering value-added services that go beyond the typical virtual assistant services. For example, you could offer consulting services, training programs, or customized packages that address specific client needs. By providing additional value to your clients, you can build stronger relationships and increase your chances of repeat business.

F. Provide excellent customer service

Customer service is key to retaining clients and building a positive reputation for your business. Respond promptly to client inquiries, be professional and courteous in all communications, and go above and beyond to exceed client expectations. Make sure your clients feel heard, appreciated, and well-supported throughout their working relationship with you.

By implementing these marketing and sales strategies, you can attract and retain high-quality clients, build a strong brand reputation, and grow your virtual assistant business.

Remember to continuously evaluate and adjust your strategies based on client feedback and market trends, and always strive to provide exceptional service to your clients.

## Operations and Management

Running a virtual assistant business involves managing multiple tasks and clients, often simultaneously. Here are some strategies to help you effectively manage your operations and provide excellent service to your clients:

A. Set clear boundaries

As a virtual assistant, it can be tempting to work around the clock to meet your client's needs. However, it is important to set clear boundaries and establish realistic work hours and expectations. Communicate your availability and response times to your clients, and be sure to take breaks and maintain a healthy work-life balance.

B. Use project management tools

Project management tools can help you stay organized and manage multiple tasks and clients efficiently. Consider using tools like Asana, Trello, or Basecamp to track tasks, deadlines, and client communication. This can help you prioritize your workload, stay on track, and avoid missed deadlines or communication gaps.

## C. Streamline your processes

To maximize your efficiency and productivity, consider streamlining your processes and workflows. For example, you could create standardized templates and checklists for common tasks, automate routine tasks using tools like Zapier or IFTTT, or outsource certain tasks to other virtual assistants or freelancers.

## D. Build strong client relationships

Building strong client relationships is key to the success of any virtual assistant business. Communicate regularly with your clients, provide regular progress updates, and ask for feedback to ensure you are meeting their needs and expectations. Consider offering additional value-added services to build stronger relationships and increase your chances of repeat business.

## E. Manage your finances

Managing your finances is an essential part of running a successful virtual assistant business. Set up a separate business bank account, track your income and expenses using accounting software like QuickBooks or Xero, and establish a pricing strategy that is competitive and reflects the value you provide. Be sure to save for taxes and set aside funds for future investments in your business.

By implementing these operations and management strategies, you can effectively manage your virtual assistant business, provide excellent service to your clients, and achieve long-term success. Remember to continuously evaluate and adjust your strategies based on client feedback and market trends, and always strive to improve your skills and knowledge as a virtual assistant.

## Challenges and Solutions

While starting and running a virtual assistant business can be rewarding, it can also come with its fair share of challenges. Here are some common challenges virtual assistants face and strategies for overcoming them:

A. Balancing multiple clients

As a virtual assistant, you may have to manage multiple clients and projects at the same time, which can be overwhelming. To address this challenge, consider setting clear boundaries and expectations with your clients, using project management tools to stay organized, and outsourcing certain tasks to other virtual assistants or freelancers.

B. Dealing with difficult clients

Dealing with difficult clients is never easy, but it is important to address their concerns and maintain a professional relationship. Consider setting clear expectations and communication protocols upfront, providing regular progress updates, and addressing any issues or concerns as soon as they arise.

C. Staying up-to-date with new technology and tools
The world of virtual assistance is constantly evolving, and it is important to stay up-to-date with new technology and tools. Consider investing in continuing education and professional development, joining online communities or networking groups, and regularly evaluating and updating your skill set.

D. Setting and managing client expectations
Setting and managing client expectations can be a challenge, but it is essential for providing excellent service and maintaining long-term client relationships. Consider establishing clear communication protocols, setting realistic deadlines and deliverables, and providing regular progress updates and status reports.

E. Finding new clients
Finding new clients can be a challenge for virtual assistants, especially when starting. Consider leveraging social media and online platforms to showcase your skills and services,

joining online communities or networking groups, and offering referral incentives to existing clients.

By addressing these common challenges and implementing strategies to overcome them, you can set yourself up for success as a virtual assistant and build a sustainable and profitable business. Remember to stay flexible and adaptable, continuously evaluate and adjust your strategies, and always prioritize providing excellent service to your clients.

## Inspiration

Becoming a successful virtual assistant is not easy, but it is possible.

There are inspiring success stories of virtual assistants who have built thriving businesses and are making a living doing what they love.

By learning from the experiences of those who are successful and taking action to build your own business, you too can achieve your goals and create a thriving virtual assistant business.

Remember to stay focused, stay motivated, and never give up on your dreams. With hard work and dedication, anything is possible.

## Conclusion

Becoming a successful virtual assistant requires hard work, dedication, and a willingness to learn and grow.

By following the steps outlined in this book, you can start and grow your own virtual assistant business and achieve your goals.

Here are some key takeaways to keep in mind:

- Start by identifying your skills, passions, and areas of expertise. This will help you determine what services to offer and how to market yourself.
- Develop a strong brand and online presence to attract clients and showcase your skills and expertise.
- Create a solid business plan and pricing structure to ensure you are charging what you're worth and providing value to your clients.
- Focus on providing excellent customer service and building strong relationships with your clients to ensure repeat business and referrals.

- Stay up-to-date on industry trends and new technologies to stay competitive and offer the best services possible.
- Don't be afraid to ask for help or seek out resources to help you grow your business.
- Remember, becoming a successful virtual assistant is a journey, not a destination.

It takes time, patience, and hard work to achieve your goals, but with the right mindset and strategies, you can build a thriving business and achieve the lifestyle and financial freedom you desire.

So go out there, take action, and start building your dream business today!